Mathematics – 1

Joseph Furnari

CALCULUS

WITHOUT LIMITS

ISBN 978-1-4452-2198-4

Printing/Distribution provided by:

Lulu Enterprises, Inc.
3131 RDU Center Dr., Ste. 210
Morrisville, NC 27560
USA
www.lulu.com
www.lulu.com/it

The book is listed on the web page
http://stores.lulu.com/giuseppefurnari,
where reviews can be posted.
Printed and distributed from lulu.com.

To my daughters
Maddalena and Marta

CALCULUS WITHOUT LIMITS

CALCULUS

WITHOUT LIMITS

The "infinitesimals", from which takes the name the infinitesimal calculus or "*the calculus*", they were grazed by the ancient Greek that never succeeded in fully them and perhaps even to understand entirely them, with theirs to confine themselves to the constructions with ruler and compasses and to the finite entities, and with the poor mastery of the continuum criterion.

If Pythagoreans and then Aristotle they described the pythagorean point as **unity endowed with position in the space**, anticipating "Cartesian" ideas, Zeno of Elea reacted like philosopher with paradoxical reasonings of strong metaphysical taste.

Archimedes, with his method of exhaustion, practically came to anticipate the integral calculus, as in his quadrature of the parabolic segment or in the volume of the segment of ellipsoid or paraboloid.

The not ended proceedings were disapproven by the Greek, and for this neither the great Archimedes reached the concept of *limit* of a function, also coming so nearby of it: rather used demonstrations with double *reductio to absurdum*.

He also reaches to calculate the surface and the volume of the sphere, after a series of theorems one of which was equivalent to integrate the *sine function*.

Apollonius of Perga in his Conics employs *straight lines of reference* as the diameter of a conic and the tangent to one extremity of his, and on them measures the distances. This is certainly equivalent to a rudimentary **analytic geometry** that anticipates of 18 centuries that Cartesian. Besides he formulates some equations of curves, even if only through oral expressions, but misses in effects a preexisting coordinate system: they is overlapped to the various curves to improve its study.

After having reached so elevated results, for the mathematics and the Greek science the decline and the obscurity it comes. For the Christians - that to be in opposition to the pagan culture they put in ridicule mathematics, astronomy and physics - was forbidden the contamination with the Greek culture. As soon as were able, they not only burned the last great collection of Greek works (300.000 manuscripts) but they acted to the same way for the whole empire attaching and murdering the pagans: the mathematician of great renown **Hypatia** was made asunder in the roads in Alexandria.

And then in 529 also in the Eastern Roman Empire all the Greek schools were closed, beginning from the Academy and from the philosophical schools in Athens. The finishing stroke was given by the conquest of Egypt from the Moslems of Omar that held in account an only book, the Koran: the Baths of Alexandria were heated burning rolls of parchment for more than six months.

Only after a millennium the mathematics and the Alexandrine sciences are been able to flourish again, and only thanks to spread some ancient works rediscoveries in some Latin translation or retranslated by Arabic interpretations.

After the contacts with the Greek culture following the Crusades, there was a great interest and different researchers were financed by Princes and Ecclesiastics to seek the most important works in Sicily, North-Africa, Spain, Middle East. Besides, were then freed from the Arabic dominion the Sicily and Toledo.

Leonard of Pisa (1170-1250), known as Fibonacci, first mathematician worthy of note in Europe, had learned the arithmetic in the Northern Africa.

Then in the Renaissance the algebra had a notable impulse and this favored the analytical geometry of the modern age, while, for instance, Apollonius of Perga (262-190 b.c.) was heavy of the algebra all geometrical of the classical age.

The Greek didn't conceive a suitable definition for the tangent to a curve in his determined point; at the most the straight line for such point was prepared in way that was not possible to trace others of it between it and the curve in matter.

But also around the tangents and the normals to the conics Apollonius it gave different theorems, in the form of theorems of maximums and minimums. These studies certainly favored those on the trajectories of the planets in the *Principia* of Newton; the Universal Gravitation was born.

The *annus mirabilis* of Newton was 1666: with the demonstration of the *Binomial Theorem*, her *Theory of the Color* that contrasted to that of Goethe, the invention of the *Infinitesimal Calculus* whose priority longly contended with Leibniz.

THE INFINITESIMAL CALCULUS

The *Infinitesimal Calculus* it consists of two important parts, both of them due to Newton and to Leibniz: the derivation and the integration. In the derivation reenters the concept of limit of a function, afterwards introduced, that has allowed to govern better the controversial infinitesimals.

The most classical problem that has given impulse to the invention of the infinitesimal calculus has been that of the tangents to a curve, to which that is equivalent of the determination of the instant speed of a point in movement.

Naturally he makes constant use of the Cartesian coordinates and the algebraic description of a curve y = f(x) on the Cartesian plane.

In the following figure is represented the typical search of the tangent to a curve in its point A seen as a chord AB when the point B is increasingly approximates to point A.

The point B" differs from the point A through two "increments" Δx and Δy, parallel to the Cartesian axes, which must tend to zero turning into the infinitesimals dx and dy.

Given that is $m_\alpha = \tan(\alpha)$ the **slope** (angular coefficient) of the tangent straight line, for the secant we will have **$m_{\beta''} = \tan(\beta'') = \Delta y/\Delta x = [f(z+h)-f(z)]/h$**.

This is called Newton's **difference quotient** (incremental ratio), and at the end, when the secant becomes tangent, we will get

$$m_\alpha = tan(\alpha) = dy/dx = f'(x).$$

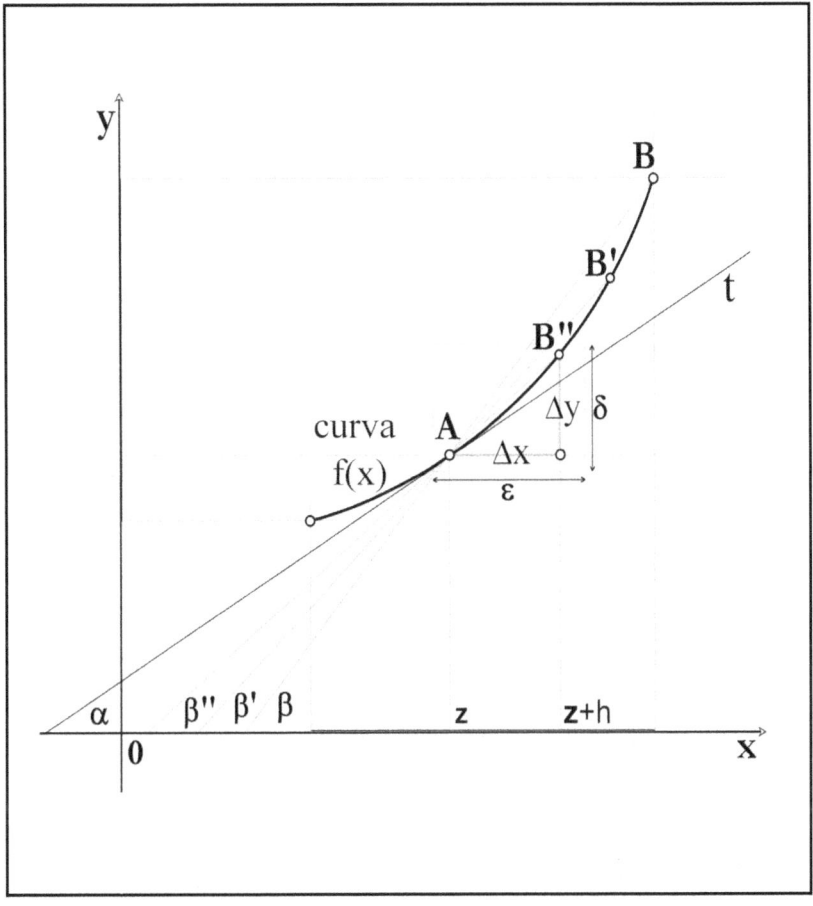

Figure 1 – Difference Quotient:

to the limit the secant is superimposed to the tangent

Nevertheless the infinitesimals **dx** and **dy**, introduced by Leibniz, they result to be the inde-terminable entities and practically unmanageables, in how much they cannot be worth zero, otherwise the ratio **dy/dx = 0/0** would not have some defined value, and cannot be different from zero, otherwise the secant cannot be identified with the tangent. Just of this type was the criticism that the bishop and philosopher Berkeley moved against the supporters of the infinitesimal calculus, that it besides perfectly worked how much mysteriously. Berkeley called they "vanishing quantities" and its criticisms were founded.

If then moves to a computational practical example, as in the case in which the function that represents the curve is $y = f(x) = x^2$, will easily have

$$m_\alpha = tan(\alpha) = \ dy/dx = [(z + dx)^2 - z^2]/dx =$$
$$= (z^2 + 2\ zdx + dx^2 - z^2)/dx =$$

$$= (2z + dx)dx/dx = 2z + dx$$

and to this point the **dx** is eliminated considering it, precisely, an infinitesimal, tending to become **dx = 0**. But only after aving "simplified" the ratio dx/dx always considering it equal to one and never tending to $\mathbf{dx/dx = 0/0}$!

This way we get the slope

$$\mathbf{m_\alpha = \tan(\alpha) = f\,'(x) = 2z}$$

or, in case the law of motion as a function of time, namely space $= s(t) = t^2$, we obtain for the velocity $\mathbf{vel = s\,'(t) = 2t}$. But bishop Berkeley's criticisms exploded more still following the operations effected with the values dx or dt, and particularly for the "reduction", as if dx or dt were ended, while later it was being eliminated considering them infinitesimals to all the effects.

More than one century needed for overcoming the problem with the approach of Karl Weierstrass that it exploited the concept of limit, just introduced by Augustin-Louis Cauchy (1760-1848), considered as approximation for improvement "as you want".

With the introduced notation, currently in use, the limit of the *difference quotients* is denoted without more to use the infinitesimals if not in the symbolism that denotes the operation of derivation $y' = f'(x) = dy/dx$. Rather, Weierstrass introduced the concept of *double limit* to narrow the area in which they found the two points A and B which identify the secant/tangent or specify the sense of instantaneous speed: they have to be within a neighbourhood ε small to like in horizontal direction (axis x or t) as well as within a neighbourhood δ small to like in vertical direction (axis y or s).

But just unfolds clearly the concept of double limit - static theory of the variable - in which doesn't appear the call to the secant that approach the tangent, it will reappear the infinitesimals! They are precisely ε and δ, which can not obviously be finished, but neither can annihilate entirely, otherwise the usual points A and B would coincide and they could not determinate any straight line.

Simply, in this way the infinitesimals ε e δ are not involved in algebraic operations nor in simpli-fications, but they *disguise* the algebraic operations and simplifications that are executed apparently on "finite increments", under the symbolism of the limit!

And yet there is a way of obtaining analytically the equation of the tangent to the curve in any point of her. So, considering the algebraic expression for her slope, you can obtain the derivative of the same curve. Equally with which is found, for instance, the tangent to a circle, namely making system of the two equations.

For instance, for the curve   ✿ $y = f(x) = x^2$, gives its points $A(z, y_0)$ [classically $A(x_0, y_0)$] and $B'(x_1, y_1)$, we will have

$$(y - y_0)/(x - z) = (y_1 - y_0)/(x_1 - z)$$

from wich $y(x_1 - z) - y_0(x_1 - z) = (y_1 - y_0)(x - z)$

or $y(x_1 - z) = y_0 x_1 - \cancel{y_0 z} + y_1 x - y_1 z - y_0 x + \cancel{y_0 z}$

and finally

$$\boxed{y = \frac{y_1 - y_0}{x_1 - z} x + \frac{y_0 x_1 - y_1 z}{x_1 - z}}$$

Where the coefficient of the x is the slope or angular coefficient of the secant AB', and in the case of the crossed spaces the instant value of the speed, while

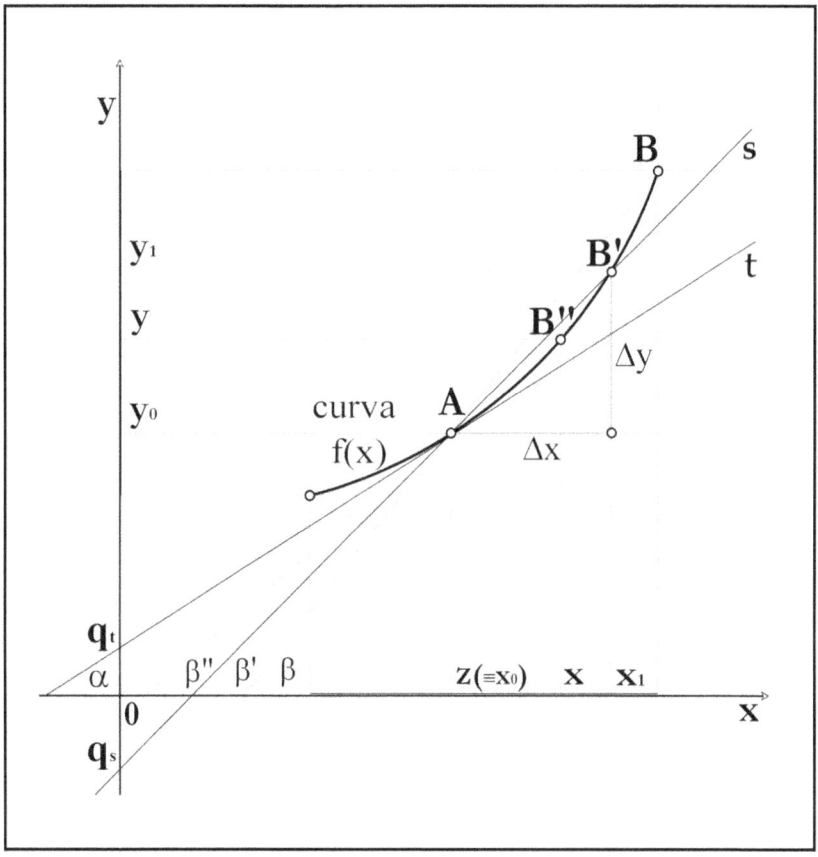

the expression of the known term is the intersection of the secant with the axis **y**, usually denoted with **q**.

You can formally write also

$$(1) \quad y = \frac{f(x_1) - f(z)}{x_1 - z} x + \frac{f(z) x_1 - f(x_1) z}{x_1 - z}$$

Such expression (1), that represents our secant for A and B', can be used **for any f(x)**.

In our specific case we replace **f(x)** with x^2 getting

$$y = (x_1^2 - z^2)/(x_1 - z) x + (z^2 x_1 - x_1^2 z)/(x_1 - z) =$$

$$= (x_1 + z)(x_1 - z)/(x_1 - z) x - x_1 z (x_1 - z)/(x_1 - z)$$

$$= (x_1 + z) x - x_1 z$$

and to this point we make to coincide B' with A replacing x_1 whit z and we get the equation of the tangent **t** :

$$y = \mathbf{2z} \, x - z^2$$

Once resolved the problem of the tangent, whose linear equation in the variable $\mathbf{x}$ is $y = 2\,z\,\mathbf{x} - z^2$ or every particular point A (z = constant), we can consider as it varies the slope of the tangent to vary some value of z, that is in correspondence of the point A that this time moves on the curve in matter.

The expression for the slope is therefore $\mathbf{m(z) = 2z}$, or rather, returning to the classical expressions of the suitable independent variable as x or as t (time, time derivative, derivative of the position, instantaneous velocity):

$$y = f(x) = x^2, \qquad y' = f'(x) = d\,f(x)\,/dx = 2\,x;$$

$$s = s(t) = t^2, \qquad vel = s'(t) = d\,s(t)\,/dt = 2\,t;$$

resulting easy to verify that, in the case the curve represents a law of the motion of a material point, the expression $\mathbf{s'(t) = d\ s(t)\ /dt = 2\ t}$ describes as it changes the instant speed of the point itself as a function of the time t.

In this way they result resolved rigorously both the problem of the tangent and of the instant speed, overcoming any contradiction and incongruity.

To better highlight what has been achieved, seems useful to consider the method of Weierstrass as all internal to the logic of Zeno when proposes his famous race between Achilles and the tortoise. When Achille reaches the position of the tortoise that had departed with an advantage, this last has already crossed a further distance that in turn Achilles will cross, but to find the tortoise always ahead to him of another stretch further, and so on to the infinity. Then Achilles, having to overcome infinite small distances, cannot reach the tortoise.

Likewise, you won't formally be able never to resolve neither the problem of the tangent neither that of the instant speed, both that try us with the infinitesimals of Newton and Leibniz or with the double limit of Weierstrass: the point B can never reach the point A; and if it reached it, they would be an only point and the tangent could not be traced neither the instant speed calculated, because it is gotten 0/0.

We know well, however, that is easy to *make system* of the two simple linear relationships for the motion to the speed of Achille and for that to the speed of the tortoise, keeping in mind of her advantage to the departure, and therefore to exactly calculate the position and the instant in which Achilles indeed reaches the tortoise, as could do whoever of us.

Exactly equally, if we set the system to take into consideration the secant line that passes for the points A and B, and we solve working substitutions that make to coincide A and B, we analytically get *the equation of the tangent* to the curve in its point A.

Extrapolating then the expression gotten for the slope of the tangent, we get, in function of the variable **x** the expression for the *instant speed,* that is finally ours *derivative*.

Doesn't stay us whether to try other cases of curves:

☆ $f(x) = x^4$

departing from (1) and replacing the $f(x)$

$$y = (x_1^4 - z^4)/(x_1 - z)\, x\ +\ (z^4 x_1 - x_1^4 z)/(x_1 - z) =$$
$$= (x_1^2 + z^2)(x_1 + z)(x_1 - z)/(x_1 - z)\, x$$
$$-\ x_1 z(x_1^3 - z^3)/(x_1 - z) =$$
$$= (x_1^2 + z^2)(x_1 + z)\, x\ -\ x_1 z(x_1^2 + x_1 z + z^2)$$

replacing then x_1 with z

$$y = (2z^2)(2z)\, x\ -\ z^2(z^2 + z^2 + z^2),$$
$$y = 4z^3\, x\ - 3z^4$$

and finally

$$y = f(x) = x^4, \qquad y' = f'(x) = d\,f(x)/dx = 4x^3;$$

$$s = s(t) = t^4, \qquad vel = s'(t) = d\,s(t)/dt = 4t^3;$$

✪ $f(x) = x^n$

departing from (1) and replacing the $f(x)$

$y = (x_1^n - z^n)/(x_1 - z)\,x + (z^n x_1 - x_1^n z)/(x_1 - z) =$

$= (x_1^{n-1} + z^{n-2} z + ...)\,(x_1 - z)/(x_1 - z)\,x$

$\qquad\qquad - x_1 z(x_1^{n-1} - z^{n-1})/(x_1 - z) =$

$= (x_1^{n-1} + z^{n-2} z + ...)\,x - x_1 z (x_1^{n-2} + x_1^{n-3} z + ...)$

replacing then x_1 with z

$y = (n\,z^{n-1})\,x - z^2\,(z^{n-2} + z^{n-2} + z^{n-2} + ...),$

$y = n\,z^{n-1}\,x - (n-1)\,z^n$

and finally

$y = f(x) = x^n, \qquad y' = f'(x) = d\,f(x)/dx = n\,x^{n-1};$

$s = s(t) = t^n, \qquad vel = s'(t) = d\,s(t)/dt = n\,t^{n-1};$

✿ $f(x) = x^{1/2}$

departing from (1) and replacing the f(x)

$$y = (x_1^{1/2} - z^{1/2})/(x_1 - z)\, \mathbf{x} \; + (z^{1/2}x_1 - x_1^{1/2}z)/(x_1 - z) =$$

$$= (x_1^{1/2} - z^{1/2})/(x_1^{1/2} + z^{1/2})(x_1^{1/2} - z^{1/2})\, \mathbf{x} \; +$$

$$+ \; x_1^{1/2}z^{1/2}(x_1^{1/2} - z^{1/2})/(x_1^{1/2} + z^{1/2})(x_1^{1/2} - z^{1/2}) =$$

$$= \; 1/(x_1^{1/2} + z^{1/2})\, \mathbf{x} \; + \; x_1^{1/2}z^{1/2}/(x_1^{1/2} + z^{1/2})$$

replacing then x_1 with z

$$y = \mathbf{1/(2\,z^{1/2})}\; \mathbf{x} \; + \; z\,(2\,z^{1/2}) \quad \text{and finally}$$

$$\boxed{\begin{array}{c} y = f(x) = \sqrt{x}, \;\; y' = \dfrac{df(x)}{dx} = \dfrac{1}{2\sqrt{x}}\,; \\[4mm] vel = s'(t) = \dfrac{1}{2\sqrt{t}}\,; \end{array}}$$

✡ **f(x) = 1/x**

departing from (1) and replacing the f(x)

$y = (1/x_1 - 1/z)/(x_1 - z)\,\mathbf{x} + [(1/z)\,x_1 - (1/x_1)\,z]/(x_1 - z) =$

$= 1/x_1z\ (z - x_1)/(x_1 - z)\,\mathbf{x}$

$+ 1/x_1z\ (x_1^{\,2} - z^2)/(x_1 - z) =$

$= -1/x_1z\ \mathbf{x} + (x_1 + z)\ /\ x_1z$

replacing then x_1 with z

$y = -\mathbf{1/z^2}\ x + 2/z$ and finally

$$y = f(x) = \frac{1}{x},\ \ y' = f'(x) = \frac{df(x)}{dx} = -\frac{1}{x^2};$$

$$vel = s'(t) = -\frac{1}{t^2};$$

✿ $f(x) = \ln_a(x)$ ✿ $f(x) = \ln(x)$

departing from (1) and replacing the f(x)

$y = [\ln_a(x_1) - \ln_a(z)]/(x_1 - z)\,\mathbf{x}\quad +$

$+ \;[\ln_a(z)\,x_1 - \ln_a(x_1)\,z]/(x_1 - z) =$

$= \ln_a(x_1/z)/(x_1 - z)\,\mathbf{x}\quad + \quad q =$

$= (z/z)\,\ln_a[(x_1/z)]^{1/(x_1 - z)}\,\mathbf{x}\quad + \quad q =$

$= (1/z)\,\ln_a[1 + (x_1 - z)/z)]^{z/(x_1 - z)}\,\mathbf{x}\quad + \quad q$

In this case we denote with q the intersection of the tangent with the y axis without calculating it, limiting us to draw the expression for it's slope m_α that is what interests us because it corresponds to the derivative that we are trying.

Besides it is necessary to remember, that the transcendent number *e* is for his own nature and definition the expression of a limit:

$$e = \lim_{n \to \infty} (1 + \frac{1}{n})^n = 2.718281828459045\ldots$$

Replacing x_1 with z we don't certainly write ∞ to the place of the exponent $[\ldots]^{z/(x_1 - z)}$ but, in this case, we can operate the passage to the limit, really because we see us the exact definition of e; therefore:

$$y = (1/z) \ln_a(e) \; x + q;$$

for $a = e$: $\qquad y = (1/z) \ln_e(e) x + q = (1/z) x + q$

and finally

$$y = f(x) = \ln_a(x), \; y' = \frac{df(x)}{dx} = \frac{1}{x} \ln_a(e), \; vel = s'(t) = \frac{1}{t} \ln_a(e);$$

$$y = f(x) = \ln_e(x), \quad y' = \frac{df(x)}{dx} = \frac{1}{x}, \quad vel = s'(t) = \frac{1}{t}.$$

APPROXIMATIONS

OF SUPERIOR ORDER

TAYLOR'S SERIES

TRIGONOMETRIC

FUNCTIONS

As had already realized the Greek affirming that among the tangent to a curve and the same curve other straight lines cannot be inserted, the equation of the tangent approach analytically in the point A that of the curve, and we can continue in the investigation both qualitatively and quantitatively.

Firstly, we say that the approximation through tangent is type linear, really because the tangent is a straight line.

As reflected in the illustration that follows, once indicated with Δx the increase of the independent variable, the corresponding increase according to the curve $f(x)$ it will be Δy; at this point we can call dy the linear increment depending on the tangent, and we can call $\varepsilon_1 \Delta x$ the difference among these two increments; intuitively - limiting us to a simple curve to monotonous trend - if Δx has the tendency to progressively decrease the same thing it will have the tendency to do ε_1, and accordingly $\varepsilon_1 \Delta x$ will have the tendency to decrease in more that linear way, that is according to an order "superior", so that altogether the segment $dy + \varepsilon_1 \Delta x$ will bring near exactly our curve $y = f(x)$.

The term dy is called ***differential*** of the function f(x), implying that it is a differential **linear.**

Then the corresponding linear differential according to the variable x is not able whether to coincide with the increment $\Delta x = dx$.

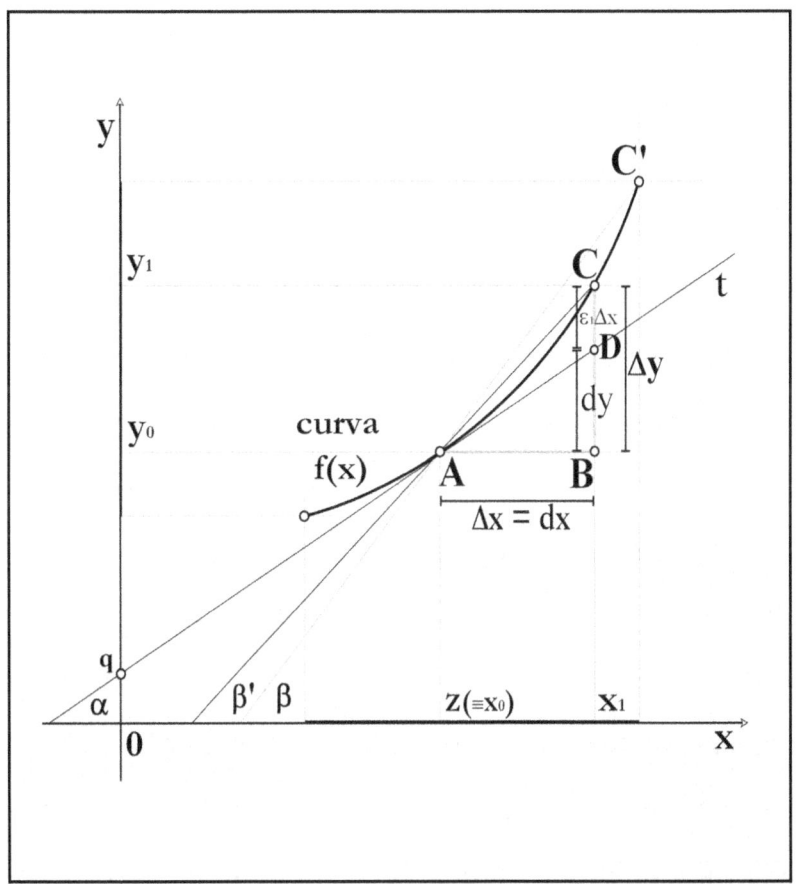

Note that there are precisely linear differentials depending on the tangent, their relationship is constant
and coincides with the derivative; that's the reason why
is current use to write

$$\dot{y} = y' = f'(x) = \frac{df(x)}{dx} = \frac{dy}{dx}$$

to indicate the derivative of the function f(x).

The differential dy and dx are typical of the
differential equations, very important in mathematical
physics. They are *functional* equations,   precisely
because the unknown to find is the same function f(x),
being known rather the relationships between his
derivative of different order and it, or between the
different derivative together.

Since the cases are very frequent in which doesn't
succeed to go up again to the solutions, you succeed in
calculating the approximate solutions really through
the differential linear dy and dx, starting from
determined initial values. The method is said of the
finite elements, and its precision depends from

how much you succeed to the increment to make small, in line with the computational skills of the used calculator.

More interesting considerations can be done when you try to seek best approximations of that linear of our tangent, to which seems to allude our increment of superior order $\varepsilon_1 \Delta x$.

If the abscissa of our point of tangency A is z, then the equation of the tangent will be

$$p_1(x) = f(z) + f'(z)(x - z)$$

where we point out the equation with $p_1(x)$ to put in evidence that deals with a first degree polynomial. You can easily verify that this polynomial in the point A[z, f(z)] coincides with the f(x), since it assumes the same values; and besides, of course, in A has the same derivative.

Likewise it is not difficult to choose a polynomial, this time of second degree, that in the same point A has in common with the f(x) the values, the first derivative and also the second derivative; it will be

$$p_2(x) = f(z) + f'(z)(x - z) + f''(z)/2!\,(x - z)^2.$$

And you can continue involving derivatives of f(x) of always highest degree, if they exist, getting the famous *formula of Taylor* - that dates back to 1715 - for the development in power series:

$$f(x) \approx p_n(x) = f(z) + f'(z)(x - z) + f''(z)/2!\,(x - z)^2 +$$

$$... + f^n(z)/n!\,(x - z)^n.$$

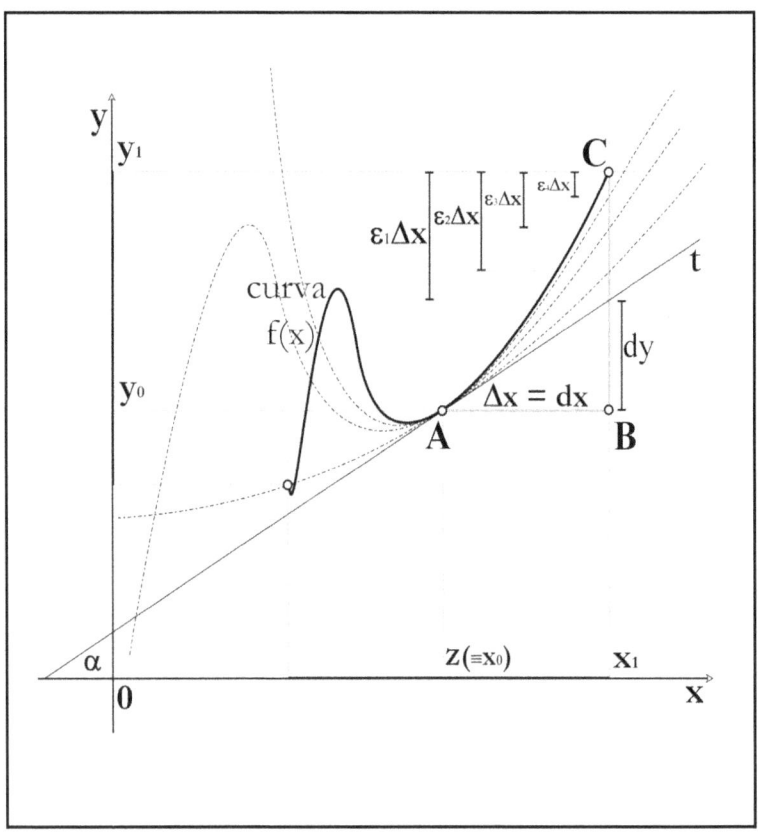

Note: here the formula of Taylor (gives in 1715) is not proved, but geometrically illustrated only.

As can be appraised by the graph, a better approximation in A is always gotten, that is of order more and more superior, getting ever more rapidly reducing increments $\varepsilon_1\Delta x$, $\varepsilon_2\Delta x$, $\varepsilon_3\Delta x$, $\varepsilon_4\Delta x$ ... $\varepsilon_n\Delta x$: the difference when z is approaching to A will decrease more quickly of $(x - z)^n$. In fact the so-called *rest*, valued in an around of the point A and not exactly in A because in such point the function f(x) and its development of Taylor in power series they necessarily coincide, is in the order of $(x - z)^n/(n + 1)!$.

Before passing to the derivatives of some trigonometric functions is opportune to examine the relationships that intervene among the values of sin(x), x   and tan(x) where x, in radians, is the argument on which are calculated the values sin(x) and tan(x), eventually using developments in series of Taylor.

In the case the circle of unit radius is considered, as in figure, x is the arc on which insist   sin(x) and tan(x).

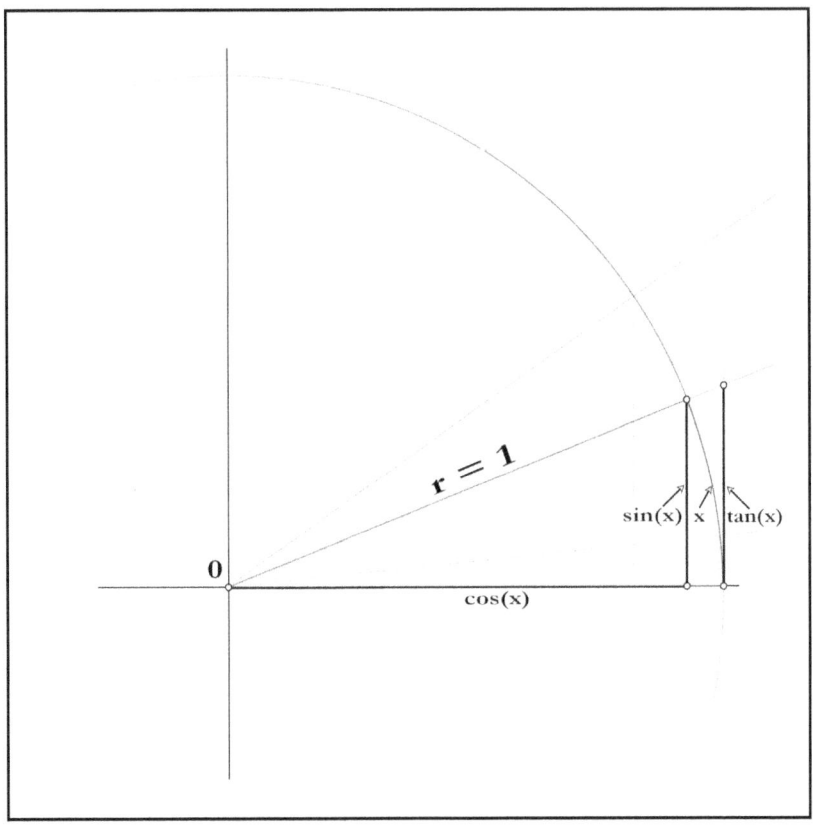

The relationship of order is immediately evident:

$\sin(x) < x < \tan(x)$,

or also

$$\frac{\sin(x)}{x} < 1 < \frac{\tan(x)}{x}, \quad \frac{\sin(x)}{x} - 1 < 0 < \frac{\tan(x)}{x} - 1;$$

Expected the evidence that for smaller values of the argument $\varepsilon = x$ are also always smaller the values

$\varepsilon_1 = \sin(x)$, $\varepsilon_2 = \tan(x)$ and $\varepsilon_3 = 1 - \cos(x)$, taking into account that $\sin(0) = \tan(0) = 0$ and $\cos(0) = 1$, can be tried to evaluate the difference $\tan(x) - \sin(x)$ simply writing:

$$\tan(x) - \sin(x) = \tan(x) [1 - \cos(x)] = \varepsilon_2 * \varepsilon_3.$$

This tells us that the difference $\tan(x) - \sin(x)$ tends to decrease with a speed "of superior order", to the point you can write $\sin(\varepsilon) \approx \varepsilon \approx \tan(\varepsilon)$ intending as "infinitely near" the meaning of the symbol " $\approx$ ", in the way of the Non-Standard analysis. It is therefore also $\sin(\varepsilon)/\varepsilon \approx 1 \approx \tan(\varepsilon)/\varepsilon$.

A confirmation is directly had considering that, for $\varepsilon > 0$, $\sin(\varepsilon) < \varepsilon < \tan(\varepsilon)$ we get immediately

$$1 < \varepsilon / \sin(\varepsilon) < 1 / \cos(\varepsilon)$$

that is $1 > \sin(\varepsilon) / \varepsilon > \cos(\varepsilon)$,

or $1 / \cos(\varepsilon) > \tan(\varepsilon) / \varepsilon > 1$;

but it is beyond doubt that $\cos(\varepsilon) \approx \cos(0) = 1$

from which inevitably **sin(ε)/ε** $\approx$ 1 and at the same

time **tan(ε)/ε** $\approx$ 1.

In the graph that follows, in which are represented both

sin(x)/x that tan(x)/x, are clearly seen that both have

the tendency to be worth   1   for small values

of the variable x.

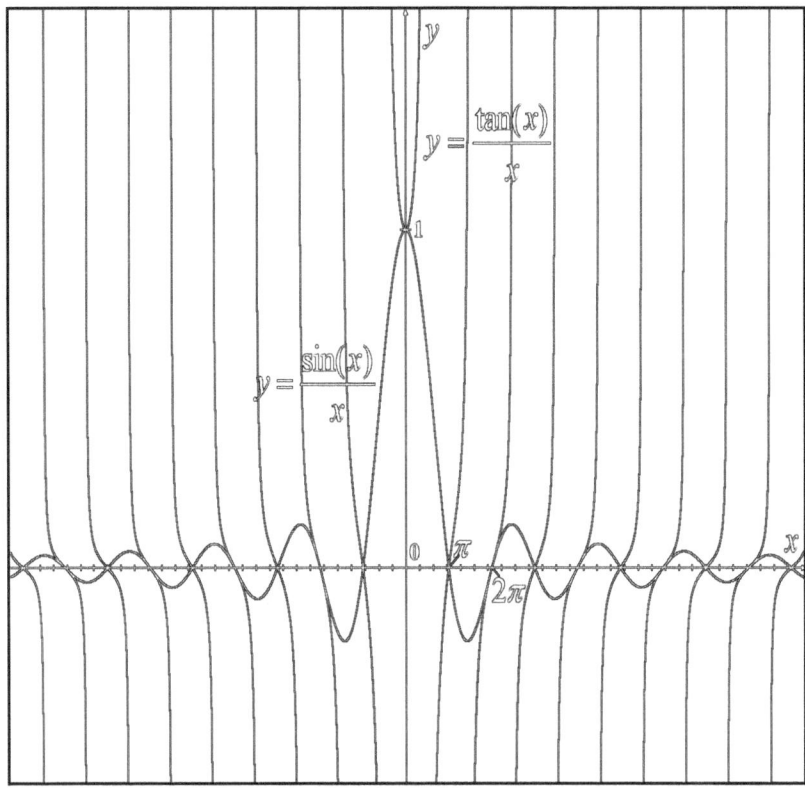

A further confirmation can be extrapolated - afterwards! - considering the developments in Taylor series for sin(x) and tan(x), that are:

$$\sin(x) = x - x^3/3! + x^5/5! - x^7/7! + \ldots$$
$$\tan(x) = x + x^3/3 + 2x^5/15 + 17x^7/315 + \ldots$$

from which

$$\sin(x)/x = 1 - x^2/3! + x^4/5! - x^6/7! + \ldots$$
$$\tan(x)/x = 1 + x^2/3 + 2x^4/15 + 17x^6/315 + \ldots$$

deducing therefore **sin(ε)/ε** ≈ 1 and **tan(ε)/ε** ≈ 1 with an approximation *of the second order*.

This is "visible" in the graphic following, where sin(x)/x and tan(x)/x are compared with the parabolas close to them $y = 1 - x^2/3!$ ed $y = 1 + x^2/3$.

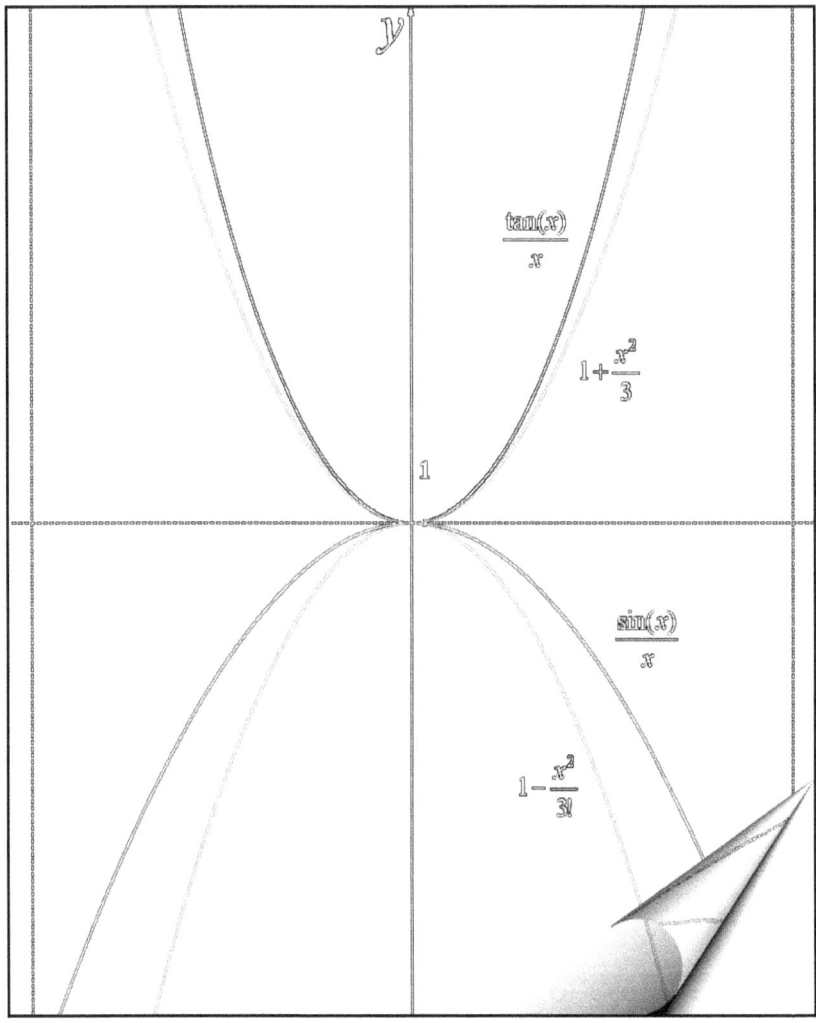

And finally

$$\tan(x) - \sin(x) = (1/3 + 1/3!)\, x^3 + (2/15 - 1/5!)x^5 +$$
$$+ (17/315 + 1/7!)x^7 + \ldots$$

that is $\mathbf{\tan(\varepsilon) - \sin(\varepsilon) \approx 0}$ with an approximation *of the third order*.

While

$$[\tan(x) - \sin(x)]/x = (1/3 + 1/3!) \, x^2 + (2/15 - 1/5!)x^4 +$$
$$+ (17/315 + 1/7!) \, x^6 + \ldots$$

that is   **$\sin(\varepsilon)/\varepsilon \approx \tan(\varepsilon)/\varepsilon \approx 1$**   always with an approximation *of the second order*.

At this point we can continue with the *trigonometric functions*:

$f(x) = \sin(x)$

always departing from (1) and replacing the f(x)

$$y = [\sin(x_1) - \sin(z)]/(x_1 - z) \, \mathbf{x} \quad +$$
$$+ [\sin(z) \, x_1 - \sin(x_1) \, z]/(x_1 - z) =$$

$$= \frac{\left\{ 2\cos\left(\dfrac{x_1 + z}{2}\right) \sin\left(\dfrac{x_1 - z}{2}\right) \right\}}{x_1 - z} x + q$$

replacing then x_1 with z

$$y = \frac{\left\{\cos\left(\frac{2z}{2}\right)\sin\frac{x_1 - z}{2}\right\}}{\frac{x_1 - z}{2}} x + q = \cos(z)\frac{\sin(\varepsilon)}{\varepsilon} x + q$$

y = **cos(z)** x + q

the reader shows that
q = sin(z) − z cos(z)

and finally

y = f(x) = sin(x), y ' = f '(x) = d f(x) /dx = cos(x),
vel = s'(t) = cos(t)

✡ **f(x) = cos(x)**

always departing from (1) and replacing the f(x)

$y = [\cos(x_1) - \cos(z)]/(x_1 - z)\,\mathbf{x} +$

$+ [\cos(z)\, x_1 - \cos(x_1)\, z]/(x_1 - z) =$

$$= \frac{\left\{2\sin\left(\frac{x_1 + z}{2}\right)\sin\left(\frac{z - x_1}{2}\right)\right\}}{x_1 - z} x + q$$

replacing then x_1 with z

$$y = \frac{\left\{ -\sin\left(\frac{2z}{2}\right)\sin\left(\frac{x_1-z}{2}\right) \right\}}{\frac{x_1-z}{2}} x + q = -\sin(z)\frac{\sin(\varepsilon)}{\varepsilon} x + q$$

then $y = -\textbf{sin (z)}\ x\ +\ q,$ and finally

$$\textbf{y = f(x) = cos(x),}\qquad \textbf{y}\,' = \textbf{d f(x) /dx} = -\textbf{sin (x),}$$

$$\textbf{vel = s}'\textbf{(t)} = -\textbf{sin (t)}$$

✿ $\textbf{f(x) = tan(x)}$

always departing from (1) and replacing the f(x)

$y = [\tan(x_1) - \tan(z)]/(x_1 - z)\,\textbf{x}\ +$

$\qquad + [\tan(z)x_1 - \tan(x_1)z]/(x_1 - z) =$

$= \{ \sin(x_1)/\cos(x_1) - \sin(z)/\cos(z) \}/(x_1 - z)\,\textbf{x}\ +\ q\ =$

$$= \frac{\left\{ \dfrac{\sin(x_1)\cos(z) - \sin(z)\cos(x_1)}{\cos(x_1)\cos(z)} \right\}}{x_1 - z} x + q =$$

$$= \frac{\left\{ \dfrac{\sin(x_1 - z)}{\cos(x_1)\cos(z)} \right\}}{x_1 - z} x + q$$

replacing then x_1 with z

$$y = 1/\cos^2(z) \left\{ \sin(x_1 - z) / (x_1 - z) \right\} x + q =$$

$$= 1/\cos^2(z) \left\{ \sin(\varepsilon) / \varepsilon \right\} x + q$$

$$y = \mathbf{1/\cos^2(z)} \; x + q = \mathbf{\sec^2(z)} \; x + q$$

and finally

$$\boxed{\begin{array}{c} y = f(x) = \tan(x), \quad y' = d\, f(x)\,/dx = 1/\cos^2(x), \\[2mm] vel = s'(t) = 1/\cos^2(t) \end{array}}$$

☼ $f(x) = \cot(x)$

always departing from (1) and replacing the $f(x)$

$$y = \frac{\cot(x_1) - \cot(z)}{x_1 - z} x + \frac{\cot(z) x_1 - \cot(x_1) z}{x_1 - z} =$$

$$= \frac{\dfrac{\cos(x_1)}{\sin(x_1)} - \dfrac{\cos(z)}{\sin(z)}}{x_1 - z} x + q =$$

$$= \frac{\dfrac{\cos(x_1)\sin(z) - \cos(z)\sin(x_1)}{\sin(x_1)\sin(z)}}{x_1 - z} x + q =$$

$$= -\frac{\dfrac{\sin(x_1)\cos(z) - \sin(z)\cos(x_1)}{\sin(x_1)\sin(z)}}{x_1 - z} x + q =$$

$$= \frac{\dfrac{\sin(x_1 - z)}{\sin(x_1)\sin(z)}}{x_1 - z} x + q$$

replacing then x_1 with z

$$y = -\frac{1}{\sin^2(z)} \frac{\sin(x_1 - z)}{x_1 - z} x + q =$$

$$-\frac{1}{\sin^2(z)} \frac{\sin(\varepsilon)}{\varepsilon} x + q$$

$$y = -1/\sin^2(z) \; x + q = -\csc^2(z) \; x + q$$

and finally

$$y = f(x) = \cot(x), \quad y' = d\,f(x)/dx = -1/\sin^2(x),$$

$$vel = s'(t) = -1/\sin^2(t).$$

HYPERBOLIC FUNCTIONS

A lso here, before passing to the derivatives of hyperbolic functions is opportune, eventually using developments in Taylor series, to examine the relationships that exits between the values of sinh(x), x and tanh(x), where x is the argument on which calcu-late the values through the exponential function e^x.

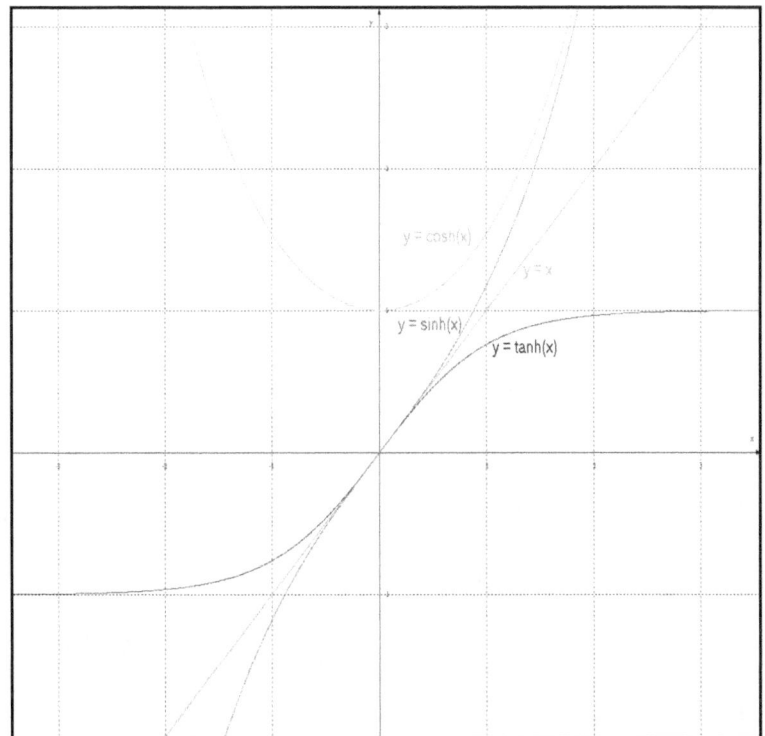

From the graph it is immediately evident the relationship of order

$$\tanh(x) < x < \sinh(x),$$

or also

$$\tanh(x)/x < 1 < \sinh(x)/x,$$

$$\tanh(x)/x - 1 < 0 < \sinh(x)/x - 1.$$

Expected the evidence that for smaller values of the argument $\varepsilon = x$ are also always smaller the values $\varepsilon_1 = \sinh(x)$, $\varepsilon_2 = \tanh(x)$ and $\varepsilon_3 = \cosh(x) - 1$, taking into account that $\sinh(0) = \tanh(0) = 0$ and $\cosh(0) = 1$, can be tried to evaluate the particular difference $\sinh(x) - \tanh(x)$ simply writing:

$$\sinh(x) - \tanh(x) = \tanh(x)\,[\cosh(x) - 1] = \varepsilon_2 * \varepsilon_3.$$

This tells us that the difference $\sinh(x) - \tanh(x)$ tends to decrease with a speed "of superior order", to the point you can write $\sinh(\varepsilon) \approx \varepsilon \approx \tanh(\varepsilon)$ intending as "infinitely near" the meaning of the symbol $" \approx "$, in the way of the Non-Standard analysis.

It is therefore also $\sinh(\varepsilon)/\varepsilon \approx 1 \approx \tanh(\varepsilon)/\varepsilon$.

A confirmation is directly had considering that, for $\varepsilon > 0$, we ave

$$\tanh(\varepsilon) < \varepsilon < \sinh(\varepsilon)$$

and we get immediately

$$1 / \cosh(\varepsilon) < \varepsilon / \sinh(\varepsilon) < 1$$

that is $\qquad \cosh(\varepsilon) > \sinh(\varepsilon) / \varepsilon > 1,$

or $\qquad 1 > \tanh(\varepsilon) / \varepsilon > 1/ \cosh(\varepsilon)$

but it is beyond doubt that $\cosh(\varepsilon) \approx 1$

from which inevitably $\mathbf{\sinh(\varepsilon)/\varepsilon} \approx 1$ and at the same time $\mathbf{\tanh(\varepsilon)/\varepsilon} \approx 1$.

In the graph that follows, in which are represented both $\sinh(x)/x$ che $\tanh(x)/x$, are clearly seen that both have the tendency to be worth 1 for small values of the variable x.

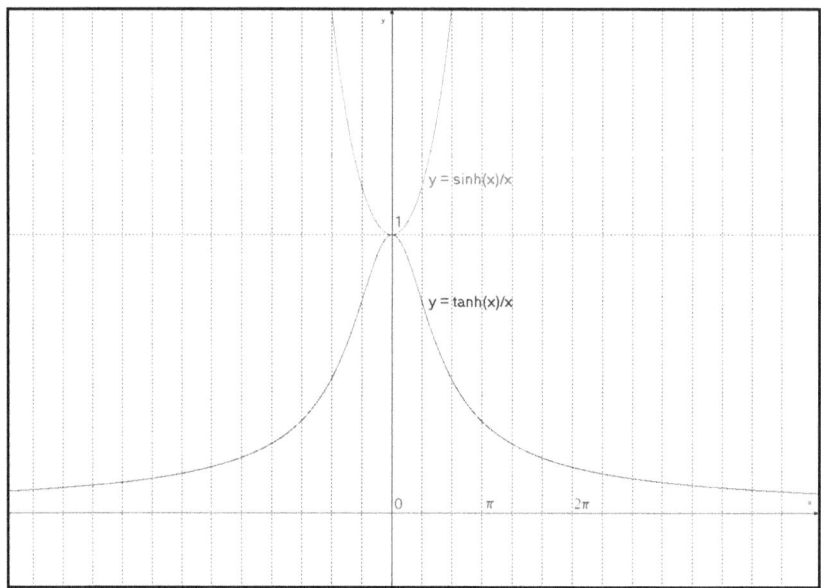

A further confirmation can be extrapolated - always afterwards - considering the developments in Taylor series for sinh(x) e tanh(x), that are:

$$\sinh(x) = x + x^3/3! + x^5/5! + x^7/7! + \ldots$$

$$\tanh(x) = x - x^3/3 + 2x^5/15 - 17x^7/315 + \ldots$$

from which

$$\sin(x)/x = 1 + x^2/3! + x^4/5! + x^6/7! + \ldots$$
$$\tan(x)/x = 1 - x^2/3 + 2x^4/15 - 17x^6/315 + \ldots$$

deducing therefore $\sinh(\varepsilon)/\varepsilon \approx 1$ and $\tanh(\varepsilon)/\varepsilon \approx 1$ with an approximation *of the second order*.

And finally

$$\sinh(x) - \tanh(x) = (1/3! + 1/3)\, x^3 +$$

$$+ (2/5! - 1/15)x^5 + (1/7! + 17/315)x^7 + \ldots$$

that is $\tan(\varepsilon) - \sin(\varepsilon) \approx 0$

with an approximation *of the third order*.

While

$$[\sinh(x) - \tanh(x)]/x = (1/3! + 1/3)\, x^2 +$$
$$+ (2/5! - 1/15)x^4 + (1/7! + 17/315)x^6 + \ldots$$

that is $\sinh(\varepsilon)/\varepsilon \approx \tanh(\varepsilon)/\varepsilon \approx 1$,

always with an approximation *of the second order*.

At this point we can continue with the *hyperbolic functions*:

✿ $f(x) = \sinh(x)$

always departing from (1) and replacing the $f(x)$

$$y = \frac{\sinh(x_1) - \sinh(z)}{x_1 - z} x + \frac{\sinh(z) x_1 - \sinh(x_1) z}{x_1 - z} =$$

$$= \frac{\left\{ 2\cosh\left(\frac{x_1 + z}{2}\right) \sinh\left(\frac{x_1 - z}{2}\right) \right\}}{x_1 - z} + q$$

replacing then x_1 with z

$$y = \frac{\left\{ \cosh\left(\frac{2z}{2}\right) \sinh\left(\frac{x_1 - z}{2}\right) \right\}}{\frac{x_1 - z}{2}} x + q =$$

$$= \cosh(z) \frac{\sinh(\varepsilon)}{\varepsilon} x + q$$

$y = \mathbf{cosh(z)}$ $\mathbf{x}$ $+ q$ and finally

$$\boxed{\begin{array}{c} \mathbf{y = f(x) = \sinh(x), \ \ y\,' = f\,'(x) = d\ f(x)/dx = \cosh(x),} \\[2mm] \mathbf{vel = s'(t) = \cosh(t)} \end{array}}$$

✿   **f(x) = cosh(x)**

always departing from (1) and replacing the f(x)

$$y = \frac{\cosh(x_1) - \cosh(z)}{x_1 - z} x + \frac{\cosh(z) x_1 - \cosh(x_1) z}{x_1 - z} =$$

$$= \frac{\left\{ 2\sinh\left(\frac{x_1 + z}{2}\right) \sinh\left(\frac{x_1 - z}{2}\right) \right\}}{x_1 - z} x + q$$

replacing then x_1 with z

$$y = \frac{\left\{ \sinh\left(\frac{2z}{2}\right) \sinh\left(\frac{x_1 - z}{2}\right) \right\}}{\frac{x_1 - z}{2}} x + q =$$

$$\sinh(z) \frac{\sinh(\varepsilon)}{\varepsilon} x + q$$

y = **sinh(z)** x + q   and finally

---

y = f(x) = cosh(x),   y' = d f(x) /dx = sinh(x),

vel = s'(t) = sinh(t)

✿   **f(x) = tanh(x)**

always departing from (1) and replacing the f(x)

$$y = \frac{\tanh(x_1) - \tanh(z)}{x_1 - z} x + \frac{\tanh(z) x_1 - \tanh(x_1) z}{x_1 - z} =$$

$$= \frac{\dfrac{\sinh(x_1)}{\cosh(x_1)} - \dfrac{\sinh(z)}{\cosh(z)}}{x_1 - z} x + q =$$

$$= \frac{\dfrac{\sinh(x_1)\cosh(z) - \cosh(x_1)\sinh(z)}{\cosh(x_1)\cosh(z)}}{x_1 - z} x + q =$$

$$= \frac{\dfrac{\sinh(x_1 - z)}{\cosh(x_1)\cosh(z)}}{x_1 - z} + q$$

replacing then x₁ with z

$$y = \frac{1}{\cosh^2(z)} \frac{\sinh(x_1 - z)}{x_1 - z} x + q =$$

$$= \frac{1}{\cosh^2(z)} \frac{\sinh(\varepsilon)}{\varepsilon} x + q$$

$$y = 1/\cosh^2(z) \ x \ + q \ = \ \text{sech}^2(z) \ x \ + \ q$$

and finally

$$y = f(x) = \tanh(x), \quad y' = d \ f(x) \ /dx = 1/\cosh^2(x),$$

$$vel = s'(t) = 1/\cosh^2(t)$$

✿ $f(x) = \coth(x)$

always departing from (1) and replacing the f(x)

$$y = \frac{\coth(x_1) - \coth(z)}{x_1 - z} x + \frac{\coth(z) x_1 - \coth(x_1) z}{x_1 - z} =$$

$$= \frac{\dfrac{\cosh(x_1)}{\sinh(x_1)} - \dfrac{\cosh(z)}{\sinh(z)}}{x_1 - z} x + q =$$

$$= \frac{\dfrac{\cosh(x_1)\sinh(z) - \cosh(z)\sinh(x_1)}{\sinh(x_1)\sinh(z)}}{x_1 - z} x + q =$$

$$= - \frac{\dfrac{\sinh(x_1)\cosh(z) - \cosh(x_1)\sinh(z)}{\sinh(x_1)\sinh(z)}}{x_1 - z} x + q =$$

$$= -\frac{\dfrac{\sinh(x_1 - z)}{\sinh(x_1)\sinh(z)}}{x_1 - z}x + q$$

replacing then x_1 with z

$$y = -\frac{1}{\sinh^2(z)}\frac{\sinh(x_1 - z)}{x_1 - z}x + q =$$

$$= -\frac{1}{\sinh^2(z)}\frac{\sinh(\varepsilon)}{\varepsilon}x + q$$

$$y = -1/\sinh^2(z)\ x\ +\ q\ =\ -\operatorname{csch}^2(z)\ x\ +\ q$$

and finally

$$y = f(x) = \cot(x), \quad y' = d\,f(x)/dx = -1/\sinh^2(x),$$

$$vel = s'(t) = -1/\sinh^2(t)$$

✿ $f(x) = e^x$

always departing from (1) and replacing the f(x)

$$y = \frac{e^{x_1} - e^z}{x_1 - z} x + q = \frac{e^{x_1}\left(1 - \dfrac{e^z}{e^{x_1}}\right)}{x_1 - z} x + q =$$

$$= e^{x_1} \frac{1 - \dfrac{1}{e^{x_1-z}}}{x_1 - z} x + q = e^{x_1} \frac{e^{z-x_1} - 1}{z - x_1} x + q$$

at this point may be account of the significant limit

$$\lim_{t \to 0} \frac{e^t - 1}{t} = 1$$

where $t = z - x_1$,   and immediately get with the
replacement of x_1 with z : $y = e^z + q$, and finally

$y = f(x) = e^x, \quad y' = d\,f(x)\,/dx = e^x, \quad vel = s'(t) = e^t$
a further demonstration for the derivative of the exponential function e^x is introduced to page 65 in the chapter on the Inverse Functions

DERIVATIVE
RULES

W hen the function y(x) to derive is, more or less simply, composed by two or more functions f(x), g(x), h(x) ... can rather easily deduce some rules of derivation that, assuming calculable the functions f '(x), g'(x), h'(x) ..., they allow to also calculate the y'(x).

For instance it is evident, that if for a date abscissa z, that of the point A, we add or we subtract the functions   f(x),   g(x),   h(x) …,   also the relative

increments Δ_f, Δ_g, Δ_h …   they are added

or subtracted. The same will happen for the

corresponding linear differentials d_f, d_g, d_h …

that directly correspond to the respective derivative.

Then we will have for the

✿   *sum:*

if $y(x) = f(x) + g(x) + h(x) …$ then

$y'(x) = d\, y(x)/dx = \{ f(x) + g(x) + h(x)…+ d_f + d_g + d_h …$

$- [f(x) + g(x) + h(x)] \}/dx =$

$= d_f/dx + d_g/dx + d_h/dx …$

that is

$$\mathbf{y'(x) = f'(x) + g'(x) + h'(x) …}$$

We will likewise have for the

✿   *difference:*

if $y(x) = f(x) - g(x)$ then

$y'(x) = d\ y(x)/dx = \{ f(x) - g(x) + d_f - d_g - [f(x) - g(x)] \}/dx = d_f/dx - d_g/dx$

that is

$$y'(x) = f'(x) - g'(x)$$

and for the

✿   *product:*

if $y(x) = f(x) * g(x)$ then

$y'(x) = d\ y(x)/dx = \{ [f(x) + d_f] * [g(x) + d_g]$

$\qquad\qquad - f(x) * g(x) \}/dx =$

$$= \{ f(x) * g(x) + f(x) * d_g + g(x) * d_f - f(x) * g(x) \}/dx =$$

$$= [f(x) * d_g + g(x) * d_f]/dx = f(x) * d_g/dx + g(x) * d_f/dx$$

that is

$$y'(x) = f(x) * g'(x) + g(x) * f'(x)$$

and, as particular case, since the derivative of a constant is zero:

✿   *product for a constant:*

if $y(x) = c * f(x)$ then

$$y'(x) = c * f'(x)$$

✿ *ratio:*

if $y(x) = f(x) / g(x)$ then

$$y'(x) = \frac{dy(x)}{dx} = \frac{\left\{\dfrac{f(x)+d_f}{g(x)+d_g} - \dfrac{f(x)}{g(x)}\right\}}{dx} =$$

$$= \frac{\left\{\dfrac{f(x)\,g(x)+g(x)\,d_f - f(x)\,g(x) - f(x)\,d_g}{\left(g(x)+d_g\right)g(x)}\right\}}{dx} =$$

$$= \frac{\left\{\dfrac{g(x)\,d_f - f(x)\,d_g}{g(x)\,g(x)}\right\}}{dx} = \frac{1}{g(x)^2}\frac{g(x)\,d_f - f(x)\,d_g}{dx} =$$

$$= \frac{1}{g(x)^2}\left[g(x)\frac{d_f}{dx} - f(x)\frac{d_g}{dx}\right]$$

that is

$$y'(x) = [\,g(x) * f'(x) - f(x) * g'(x)\,] / g(x)^2$$

and, as particular case, since the derivative of a constant is zero:

✦   *division for a constant:*

if $y(x) = f(x) / c$ then

$$y'(x) \; = \; c * f'(x) / c^2 \; = \; f'(x) / c$$

finally, for a function of function, we will have:

✦   *composite function:*

if $y(x) = f[\, g(x)\,]$ then

$$y'(x) = \frac{dy(x)}{dx} = \frac{dy(x)}{dg(x)} \frac{dg(x)}{dx} = \frac{df[g(x)]}{dg(x)} \frac{dg(x)}{dx}$$

that is

$$y'(x) \; = \; f'(x)_{dg(x)} \; * \; g'(x)_{dx}$$

and, in the case of functions of functions in succession (chain rule)

$$y(x) = f\left\{g\left[h(x)\right]\right\}$$

$$y'(x) = f'(x)_{dg[h(x)]} * g'(x)_{h(x)} * h'(x)_{dx}$$

INVERSE FUNCTIONS

$\textbf{A}$ special case of rules of derivation is that related to the inverse functions, i.e. in the case is had $y = f(x)$ in exact correspondence of $x = f^{-1}(y)$.

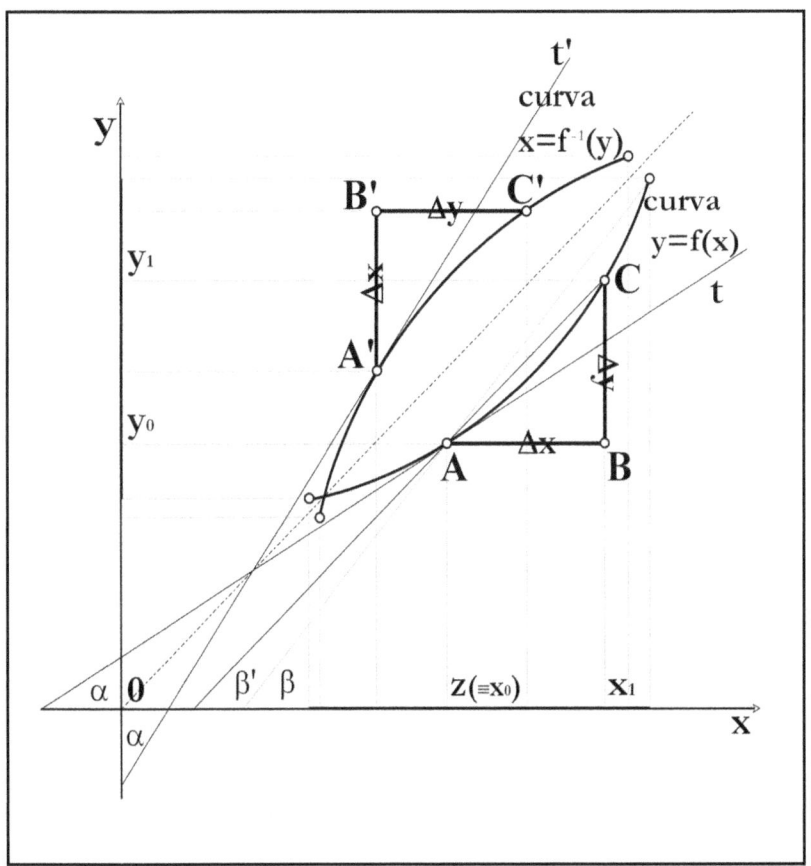

It is evident that they will result exchanged the increments Δx and Δy, as also the linear differentials dx and dy, and accordingly we will have

$$y'(x) = \frac{dy(x)}{dx} = \frac{1}{\dfrac{dx}{dy(x)}} = \frac{1}{x'(y)}$$

with $x(y) = f^{-1}(y)$

in other words

$$y'(x) = [\, x\,'(y)_{dy}]^{-1} = 1/f^{-1}{}'(y)_{dy}$$

Through the rule of the derivation of an inverse function we will obtain some further derivatives.

✿ $f(x) = a^x$,   ✿ $f(x) = e^x$

Applying the rule of the derivation of an inverse function, that in this case is $x = \ln_a(y)$, we get

$$f'(x) \;=\; [x'(y)_{dy}]^{-1} \;=\; 1/f^{-1\prime}(y)_{dy} \;=$$

$$=\; 1\,/\,[\, dx\,/\,d\,y(x)\,] \;=\; 1/[d\,\ln_a(y)/dy] =$$

$$=\; 1/[(1/y)\,\ln_a(e)] = y\,\ln_e(a) = a^x\,\ln_e(a)$$

i.e.

$d\,a^x/dx = a^x\,\ln_e(a)$ and in the case $a = e$:

$$d\,e^x/dx = e^x$$

✡ **f(x) = asin(x)**

Applying the rule of the derivation of an inverse function, that in this case is x = sin(y), we get

$$f'(x) = [x'(y)_{dy}]^{-1} = 1/f^{-1\prime}(y)_{dy} =$$

$$= 1/[d\sin(y)/dy] = 1/\cos(y) =$$

$$= 1/(1-\sin^2(y))^{1/2} = 1/(1-x^2)^{1/2}$$

or

$$y' = \frac{d[a\sin(x)]}{dx} = \frac{1}{\sqrt{1-x^2}},$$

$$vel = s'(t) = \frac{1}{\sqrt{1-t^2}}$$

✿　　　　**f(x) = acos(x)**

Applying the rule of the derivation of an inverse function, that in this case is $x = \cos(y)$, we get

$$f'(x) = [x'(y)_{dy}]^{-1} = 1/f^{-1}{}'(y)_{dy} =$$

$$= 1/[d\cos(y)/dy] = -1/\sin(y) =$$

$$= -1/(1-\cos^2(y))^{1/2} = -1/(1-x^2)^{1/2}$$

that is

$$y' = \frac{d[a\cos(x)]}{dx} = -\frac{1}{\sqrt{1-x^2}},$$

$$vel = s'(t) = -\frac{1}{\sqrt{1-t^2}}$$

✿   **f(x) = atan(x)**

Applying the rule of the derivation of an inverse
function, that in this case is x = tan(y), we get

$$f'(x) = [x'(y)_{dy}]^{-1} = 1/f^{-1'}(y)_{dy} =$$

$$= 1/[d \tan(y)/dy] = 1/\sec^2(y) = \cos^2(y)$$

$$= 1/(1 + \tan^2(y)) = 1/(1 + x^2)$$

or

$$y' = \frac{d[a\tan(x)]}{dx} = \frac{1}{1 + x^2},$$

$$vel = s'(t) = \frac{1}{1 + t^2}$$

✡ f(x) = asinh(x)

Applying the rule of the derivation of an inverse function, that in this case is x = sinh(y), we get

$$f'(x) = [x'(y)_{dy}]^{-1} = 1/f^{-1\prime}(y)_{dy} =$$

$$= 1/[d\ \sinh(y)/dy] = 1/\cosh(y) =$$

$$= 1/(1+\sinh^2(y))^{1/2} = 1/(1+x^2)^{1/2}$$

that is

$$y' = \frac{d[a\sinh(x)]}{dx} = \frac{1}{\sqrt{1+x^2}},$$

$$vel = s'(t) = \frac{1}{\sqrt{1+t^2}}$$

✿ $f(x) = acosh(x)$

Applying the rule of the derivation of an inverse function, that in this case is $x = cosh(y)$, we get

$$f'(x) = [x'(y)_{dy}]^{-1} = 1/f^{-1'}(y)_{dy} =$$

$$= 1/[d \cosh(y)/dy] = 1/\sinh(y) =$$

$$= \pm 1/(\cosh^2(y) - 1)^{1/2} = \pm 1/(x^2 - 1)^{1/2}$$

or

$$y' = \frac{d[a\cosh(x)]}{dx} = \pm\frac{1}{\sqrt{x^2 - 1}},$$

$$vel = s'(t) = \pm\frac{1}{\sqrt{t^2 - 1}}$$

✿   **f(x) = atanh(x)**

Applying the rule of the derivation of an inverse function, that in this case is   x = tan(y), we get

$$f'(x) \; = \; [x'(y)_{dy}]^{-1} \; = \; 1/f^{-1}{}'(y)_{dy} \; =$$

$$= \; 1/[d \; \tanh(y)/dy] = \; 1/\mathrm{sech}^2(y) = \cosh^2(y) \; =$$

$$= \; 1/(1 - \tan^2(y)) \; = \; 1/(1 - x^2)$$

or

$$y' = \frac{d[a\tanh(x)]}{dx} = \frac{1}{1 - x^2},$$

$$vel = s'(t) = \frac{1}{1 - t^2}.$$

THE INFINITESIMAL
CALCULUS

L eibniz affirmed that should not be underestimated too much the results reached in the antiquity: "understanding Archimedes and Apollonius, will be less admired the results subsequently reached by the most eminent mathematicians".

And in effects, for example, it is known that Apollonius had to be able to determine a conic through five points, but doesn't speak about this in his *Conics*.

It will be then an important argument in the Newton's *Principia*.

It is possible however that Apollonius spoke in the Book VIII gone lost, as generically had anticipated in the preface to the Book VII. Unfortunately, thanks to Christians and Ottomans, or to the most greater monotheisms, big part of the ancient mathematics has gone lost.

As it regards the secular matter of the infinitesimal quantities, a very ancient trace goes up again to Euclid; he treats, in the Book III of its famous ***Elements***, of the **Proposition 16**: ***The straight line drawn to right angle to the end of the diameter of a circle, stays out of the circle, and in the inclusive space between the line and the circumference cannot be interpose any other straight line; besides the angle of the semicircle is greater, and the remaining angle smaller, of every acute rectilinear angle.***

For the first time a not rectilinear angle is taken into account: the "***remaining angle***", that Euclid doesn't expressly consider null but "***smaller of every acute rectilinear angle***", it has a curvilinear side, being an arc of circumference.

For its characteristic shape the Greeks have called it "**horn angle**".

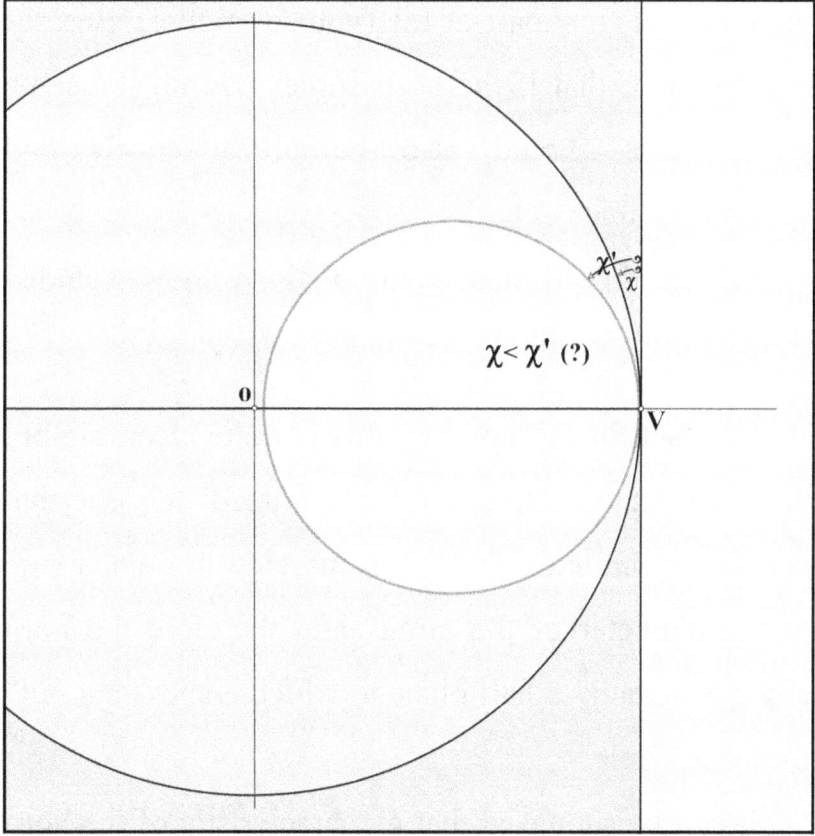

There has classically been already attention on this particular angle, and Proclus speaks of it as of a real angle. Its width became a matter that was also debated in the late Middle Ages and in the Renaissance. They were interested of it: Cardano, Peletier, Vieta, Galilco, Wallis and others, that were perplexed for the fact that tracing a smaller circle the width of the horn angle should increase, because the whole is greater of one part of his (you see figure). Besides was thought, that if the horn angles had all the same width zero, should also be all identicals and superimposables.

Others, due to these contradictions, they excluded that the surface called "horn angle" was an angle.

And in fact, in my opinion, this is quite clear: simply draw a relatively small circle, to immediately see that cannot be limited only us on one side in comparison to the diameter of the circle, and therefore the horn angle is actually a half plane to which comes "cut out" the same circle.

It can be supposed that a characteristic of the horn angles it is that they introduce a certain "scale invariance", and that from this point of view all the horn angles are similar between them. On the other hand, the relationship between the area of a half plane and that of a circle of finite radius is a relationship of the type infinite/finite, indifferent to the particular finite value of the area of the circle.

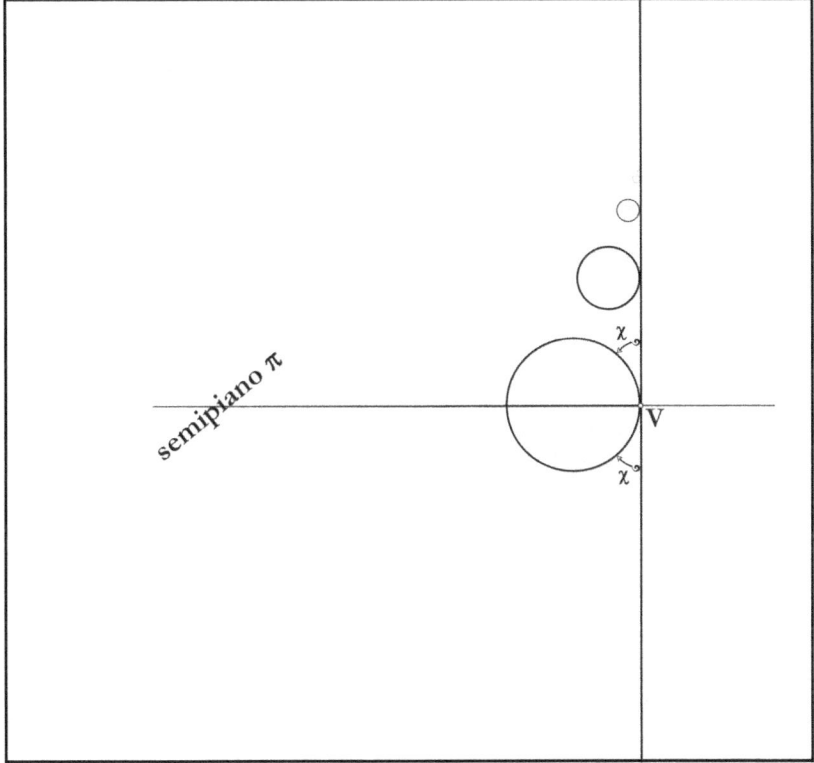

Another consideration that goes beyond the classical methodologies can be express referring to the Analysis Not-standard introduced by Abraham Robinson (1919-1974) that provided a logical status suitable to the infinitesimals of Leibniz using sophisticated techniques of theory of the models. On each straight line that delimits a half plane, or on every regular curve (with continuous derivative) that delimits an area of the plane, you can imagine that they insist infinite circles of infinitesimal radius each with its horn angles; indeed the half plane or the area could be delimited rather than from points, from infinite circles of infinitesimal radius, every practically indistinguishable from a geometric point. The same rectilinear or curved lines can think as geometrical loci either compounds from infinite points or infinite circles of infinitesimal radius…

We skip these "esotericisms". Having gotten here directly and exactly the derivative, operating the opportune substitutions, it is not necessary to make reference to the analysis Not-standard to face the problem of the infinitesimals anymore, at least in elementary limits. Currently us, that have the fortune to take advantage of the infinitesimal calculus, and we have familiarity with methods that use the so-called *difference quotient* and with the chord that tends to approach the tangent in a determined point of a curvilinear arc, we immediately should note that Euclid with its Proposition 16 have come upon in the infinitesimal quantities, and have treated them with correctness and great mastery.

Everything this becomes still more evident if we consider the following illustration, where a point B moves along the arc that delimits the horn angle,

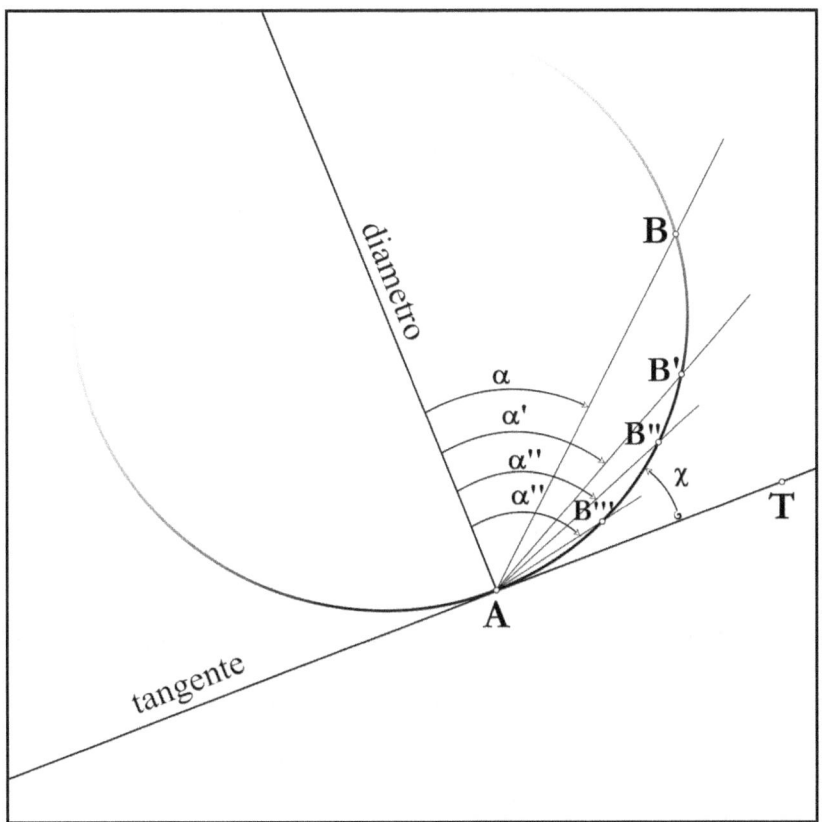

passing for B', B"…, approaching itself to the point A that is its vertex. Joining the points B, B', B "… with the vertex A are gotten some chords that gradually approach to the tangent AT. While the acute angles betwen the chords and the diameter have the tendency to become a right angle, without never reaching it, the width of the acute angles between the chords and the tangent becomes more and more small. Everything this must to have seen Euclid, and it is not very distant from the fulcrum of the infinitesimal calculus. Finally it is notable as Euclid, contrarily of Leibniz that for long time has held the infinitesimal extremely small but finite and constant, is not fallen into any error. He would have canned to affirm that the width of the horn angle is very small, or zero; instead has abstained, limiting himself to affirm that it is smaller than any acute rectilinear angle.

Naturally something remained pending, as necessarily something remains pending with the most recent concept of infinitesimal quantity: it is somehow absurd, an inferior limit to the infinitely small numbers not existing, at least in the Archimedean fields. For the standard analysis the infinitesimals would only be useful inventions.

It is generally considered that Weierstrass has overcome the problem of the infinitesimals restraining the convergence of the values of the functions with his method of the double limit, the so-called static theory of the variable. In reality I believe that the double limit only disguises the infinitesimals, as already I said on page 12.

The same Weierstrass has introduced examples of continuous functions that don't admit derivative in none of their points, and has reached the point to conclude that the class of the continuous functions is notably ampler than those derivable.

Finally, you can take the example of some simple functions. For instance

$$y = \begin{cases} x\sin\left(\dfrac{1}{x}\right) & per\ x \neq 0 \\ 0 & per\ x = 0 \end{cases}$$

that in its point (0,0) it doesn't admit derivative, also being possible to indefinitely confine her within neighbourhoods ε and δ small to like, then just with the method of the double limit of Weierstrass.

It follows the graph and the particular with the neighbourhoods ε e δ.

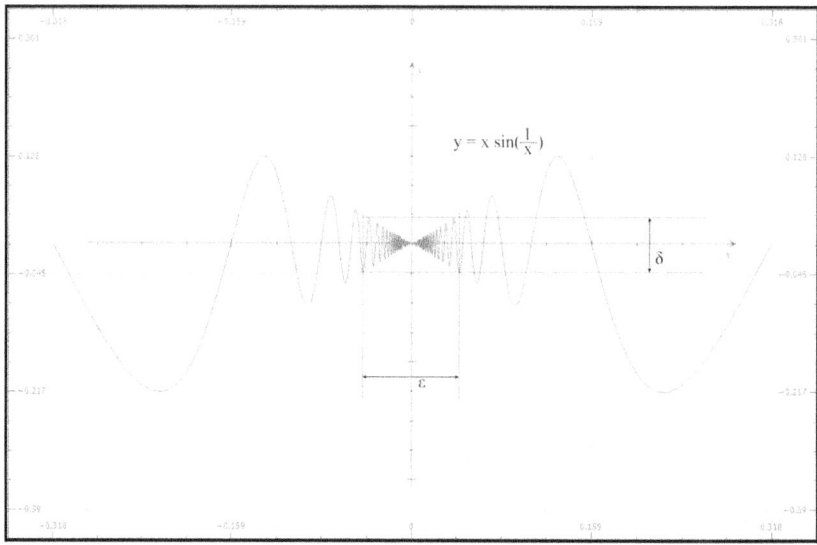

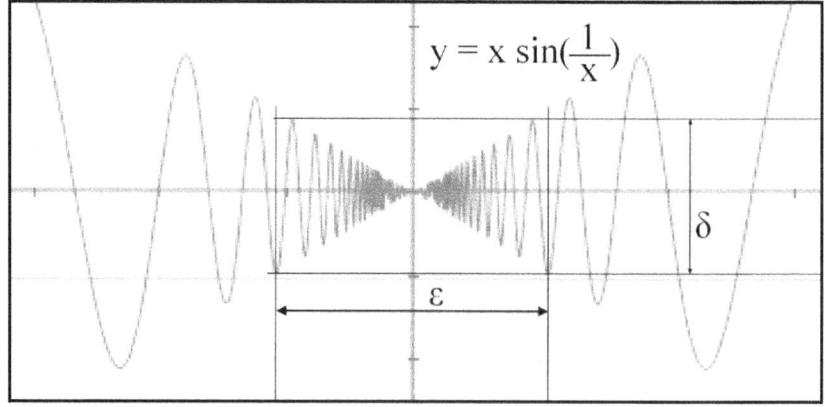

In other words, the double limit of Weiertrass doesn't guarantee neither the existence of the point of the function around which "tighten" the infinitesimal neighbourhoods ε and δ, neither the existence of the derivative in that point.

And this is worth both for banal functions as $y = |x|$, as for the particular functions introduced by the same Weierstrass that also continuous anywhere they are not derivable in any point.

An example is

$$y = -\sum_{n=0}^{7} \left(\frac{2}{3}\right)^{n} \sin(2^{n}x)$$

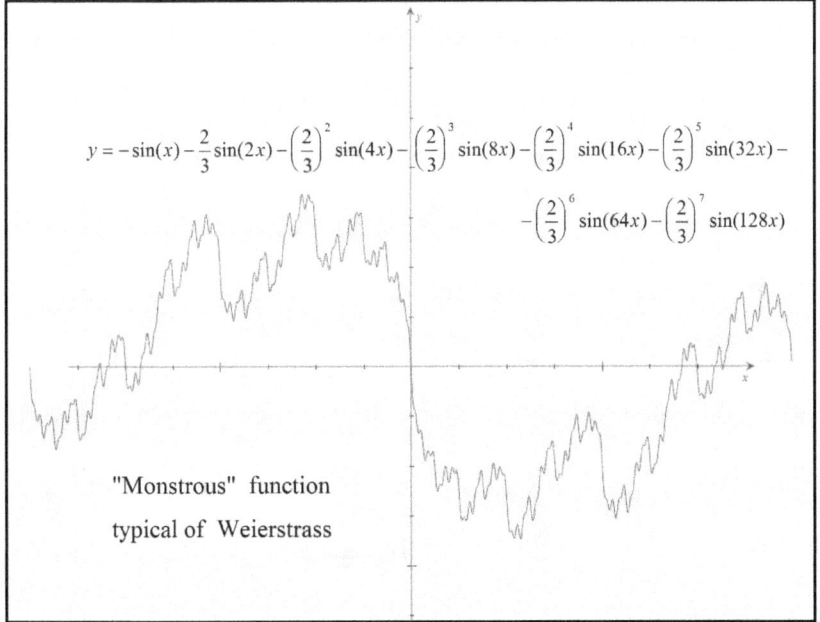

$$y = -\sin(x) - \frac{2}{3}\sin(2x) - \left(\frac{2}{3}\right)^{2}\sin(4x) - \left(\frac{2}{3}\right)^{3}\sin(8x) - \left(\frac{2}{3}\right)^{4}\sin(16x) - \left(\frac{2}{3}\right)^{5}\sin(32x) -$$

$$-\left(\frac{2}{3}\right)^{6}\sin(64x) - \left(\frac{2}{3}\right)^{7}\sin(128x)$$

"Monstrous" function
typical of Weierstrass

In other words

$$y = -\sin(x) - \frac{2}{3}\sin(2x) - \left(\frac{2}{3}\right)^2 \sin(4x) -$$

$$- \left(\frac{2}{3}\right)^3 \sin(8x) - \left(\frac{2}{3}\right)^4 \sin(16x) - \left(\frac{2}{3}\right)^5 \sin(32x) -$$

$$- \left(\frac{2}{3}\right)^6 \sin(64x) - \left(\frac{2}{3}\right)^7 \sin(128x)$$

what, continuing for $n \to \infty$, that is

$$y = -\sum_{n=0}^{\infty} \left(\frac{2}{3}\right)^n \sin(2^n x)$$

it becomes infinitely jagged to every scale and enlargement, so much to be been listed among the fractals. Then, also being continuous, it is not considered derivable in none of her points. We will see however that it is not exactly so, in one next job of mine.

Nevertheless, it doesn't need to think that this type of functions is so irremediably strange and esoteric, even if fluently are defined as "monstrous". You are enough to think that also the normal and smooth function $y = \sin(x)$ can be written and graphically represented in the form of her development of Taylor in power series.

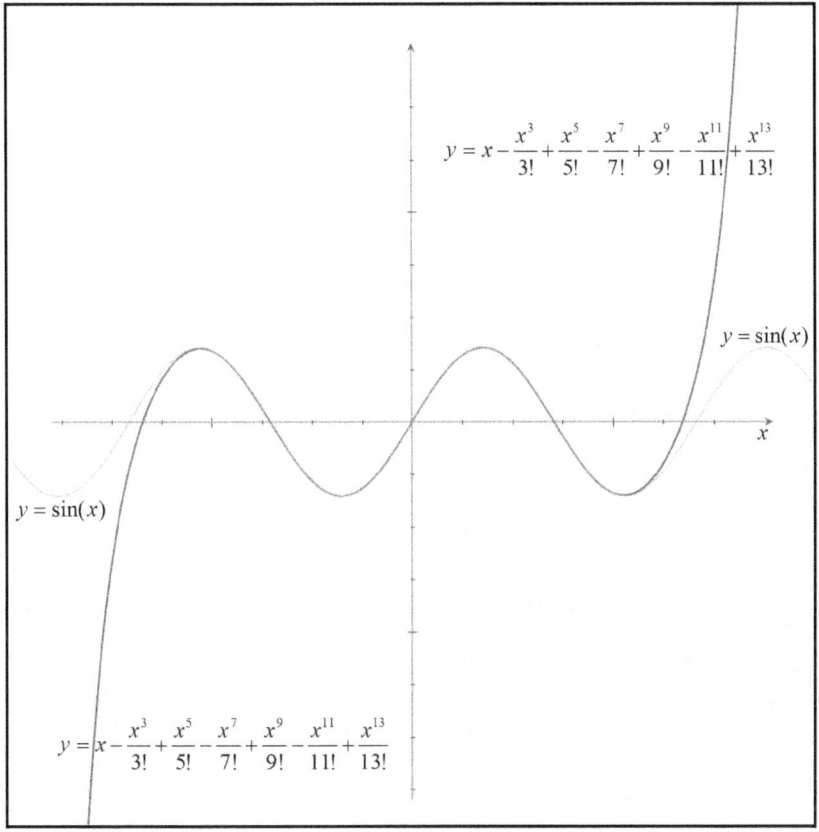

$$y = x - \frac{x^3}{3!} + \frac{x^5}{5!} - \frac{x^7}{7!} + \frac{x^9}{9!} - \frac{x^{11}}{11!} + \frac{x^{13}}{13!}$$

$$y = \sin(x)$$

$$y = \sin(x)$$

$$y = x - \frac{x^3}{3!} + \frac{x^5}{5!} - \frac{x^7}{7!} + \frac{x^9}{9!} - \frac{x^{11}}{11!} + \frac{x^{13}}{13!}$$

In the graph here above they are compared the sine function and her truncate development to only 7 terms: can easily realize as the "monstrous" development with all of its infinite terms simply will coincide with the sine function.

$$y = \sin(x) = x - \frac{x^3}{3!} + \frac{x^5}{5!} - \frac{x^7}{7!} + \frac{x^9}{9!} - \frac{x^{11}}{11!} + \frac{x^{13}}{13!} - \frac{x^{15}}{15!} + \ldots$$

Concluding, it is opportune to avoid the more possible the evanescent infinitesimals that bring to logical contradictions and they don't succeed in being satisfactory in the practical examples.

Naturally the concept of limit introduced by Cauchy results very useful. As the analysis of the relations among x, sin(x) and tan(x), for small values of the variable x, succeeding in checking also sin(x)/x and tan(x)/x.

And the concept of limit is not certain avoidable. Just think that is even used to define the value of a very important constant as the Euler's number *e*.

What counts in the infinitesimal calculus - by now cannot be done to less than call it so - it is not to avoid to implicate the limits at all costs: it is not reason of it. It is important not to use the concept of limit, and therefore that of infinitesimal, directly in the mechanism proper of the derivation; for instance performing the opportune substitutions that ***exactly*** make to ***coincide*** the point B with the point A, without instead having to approach indefinitely, remaining embroiled in the considerations on the infinitesimal distances.

I think I have succeeded.

from Pinerolo (Turin) Italy − October 2007

analytical index

Series "*mathematics*"

1 – **Calculus without limits**

will follow

– **Functions, limits, continuity**

– **Three articles announcing a mystery**

– **The mystery of the fifth postulate**

– **Complex transformations,**
  Oval polynomial and polynomials

– **Circles, hypercircles and conic on the complex plane**

– **Differential equations**

annotations

www.ingramcontent.com/pod-product-compliance
Lightning Source LLC
Chambersburg PA
CBHW071236170526
45165CB00003B/1121